## cloverleaf books™
### Space Adventures

# To the Sun!

Jodie Shepherd
Illustrated by Paula Becker

M MILLBROOK PRESS • MINNEAPOLIS

Text and illustrations copyright © 2017 by
Lerner Publishing Group, Inc.

All rights reserved. International copyright secured. No part of this book may be reproduced, stored in a retrieval system, or transmitted in any form or by any means—electronic, mechanical, photocopying, recording, or otherwise—without the prior written permission of Lerner Publishing Group, Inc., except for the inclusion of brief quotations in an acknowledged review.

Millbrook Press
A division of Lerner Publishing Group, Inc.
241 First Avenue North
Minneapolis, MN 55401 USA

For reading levels and more information, look up this title at www.lernerbooks.com.

Main body text set in Slappy Inline 22/28.
Typeface provided by T26.

**Library of Congress Cataloging-in-Publication Data**

Names: Shepherd, Jodie, author. | Becker, Paula, 1958– illustrator.
Title: To the Sun! / by Jodie Shepherd ; illustrated by Paula Becker.
Description: Minneapolis : Millbrook Press, [2017] | Series: Cloverleaf Books. Space Adventures | Audience: Ages 5–8. | Audience: K to grade 3. | Includes bibliographical references and index.
Identifiers: LCCN 2016009648 (print) | LCCN 2016014857 (ebook) | ISBN 9781512425383 (lb : alk. paper) | ISBN 9781512428353 (eb pdf)
Subjects: LCSH: Sun—Juvenile literature.
Classification: LCC QB521.5 .S484 2017 (print) | LCC QB521.5 (ebook) | DDC 523.7--dc23

LC record available at https://lccn.loc.gov/2016009648

Manufactured in the United States of America
1-41306-23250-4/12/2016

# TABLE OF CONTENTS

**Chapter One**
Where Is the Sun?.....4

**Chapter Two**
Circling the Sun.....8

**Chapter Three**
Watch Out!.....14

**Chapter Four**
The Sun Returns!.....20

Way to Grow!....22

Sun Safety Tips....22

Glossary....23

To Learn More....24

Index....24

## Chapter One
# Where Is the Sun?

Yawn. Leela loves science, but she feels so sleepy. She looks out the window. It's cloudy and raining for the third day in a row. Where is the sun?

"The sun is the closest star to Earth," Mr. Holt continues. "But it's still pretty far away. A jet plane would take about twenty years to fly there."

*No way!* thinks Leela. *I bet I could get there faster. And I could bring home enough sunshine to give everybody energy.*

Venus
Earth
Mercury
Jupiter
Mars
Saturn
Uranus
Neptune

The sun is 93 million miles (150 million kilometers) from Earth.

A sunbeam shines on Leela's shoulder. The ground begins to shake. And just like that, *whoosh!* She's off on a supersonic spacecraft!

Leela looks at the sun's surface. It's not solid like Earth's.

Earth and the solar system's other planets orbit, or travel around, the sun. The sun's gravity keeps the planets from floating away.

Uh-oh, trouble! Things are getting really bright and *really* hot, even inside the spacecraft.

"Too hot!" Leela shouts. "It's a good thing we're not usually this close to the sun."

The ancient Greeks believed that their sun god, Apollo, pulled the sun across the sky every day in his chariot.

Leela needs sunshine to wake up her class, but not *this* much of it. She had better do something—fast!

## Chapter Three
# Watch Out!

Just in time, Leela finds a pair of superthick sunglasses and a special heatproof outfit. "What makes the sun so hot?" she wonders out loud.

"The sun is mostly made of hydrogen, and it is always burning," the computer says. "It's like explosion after explosion for billions of years."

"Without the sun, there would be no night and no day," the computer tells Leela.

"No winter, spring, summer, or fall," Leela replies, "and no food to eat. Yikes!"

Solar flares are eruptions on the sun that release huge amounts of energy and heat into space.

"Here comes another burst!" yells Leela. "This spacecraft is filled top to bottom with sunlight. Let's get this baby home!"

## Chapter Four
# The Sun Returns!

"Look! The sun's out!" Leela hears a classmate shout.

Leela stops daydreaming. There are sunbeams everywhere. "I did it!" she shouts.

She tells her class all about her adventure. "We can't live *on* the sun," she says. "But we sure can't live without it."

# Way to Grow!

Humans aren't the only ones who get energy from the sun. So do plants! This experiment will show you how sunlight can help a plant grow.

### What You Will Need

two of the same type of small potted plants
a sunny location, like a windowsill or porch
a dark location, like a closet or basement

### How to Do Your Experiment

1) Put one plant in the sunny location.

2) Put the second plant in the dark location.

3) Keep them watered for a week or two. Make sure the soil is moist, but not wet.

**Which plant looks healthier? Why do you think that is?**

# Sun Safety Tips

- **Never look directly at the sun.**

- **Wear sunscreen, sunglasses, and a hat to protect against the sun's rays.**

- **Keep yourself cool by drinking plenty of water.**

# GLOSSARY

**core:** the very middle of something

**eruption:** a sudden and violent explosion that throws material into space

**galaxy:** a group of millions or billions of stars, held together by gravity. Earth is in the Milky Way galaxy.

**gravity:** the force that pulls one object toward another

**hydrogen:** a gas that burns easily

**orbit:** to travel around something on a path, as Earth travels around the sun

**planet:** a body in space that is in orbit around the sun

**solar system:** the sun and the eight planets and their moons that travel around our solar system as well as smaller bodies, such as comets, meteoroids, and asteroids

**supersonic:** faster than the speed of sound

# TO LEARN MORE

## BOOKS

**Hughes, Catherine D.** *First Big Book of Space.* Washington, DC: National Geographic, 2012.
Learn more information about the sun, our solar system, and the rest of the universe through the beautiful photos in this book!

**Jemison, Mae, and Dana Meachen Rau.** *Exploring Our Sun.* New York: Children's Press, 2013.
Written by former astronaut Dr. Mae Jemison, this book is all about exploration of the sun and how scientists are preparing to travel closer to our nearest star.

**Waxman, Laura Hamilton.** *The Sun.* Minneapolis: Lerner Publications, 2010.
Check out this book for a complete, up-close introduction to the sun and its importance in our solar system.

## WEBSITES

**NASA Kids' Club**
http://www.nasa.gov/audience/forkids/kidsclub/flash/index.html
Get the latest news and videos, including mission updates, from the NASA space agency.

**Sun Safety Alliance**
http://www.sunsafetyalliance.org/kids.html
Visit this website to learn sun safety tips and check out fun sun safety activities!

**Sun—the Closest Star to Our Earth**
http://easyscienceforkids.com/all-about-the-sun
Watch videos, read answers to other kids' questions, and learn more fun facts about the sun on this website.

**LERNER SOURCE™**
Expand learning beyond the printed book. Download free, complementary educational resources for this book from our website, www.lernerresource.com.

# INDEX

energy, 5–6, 19, 21

light, 5, 19

orbiting, 10–11

solar flares, 19

solar system, 10

stars, 5–6, 9